CHEMISTRY COLORING BOOK

BY KID KONGO

Copyright © 2016 Kid Kongo

All rights reserved.

ISBN: 10:1532874243
ISBN-13: 978-1532874246

I0482147

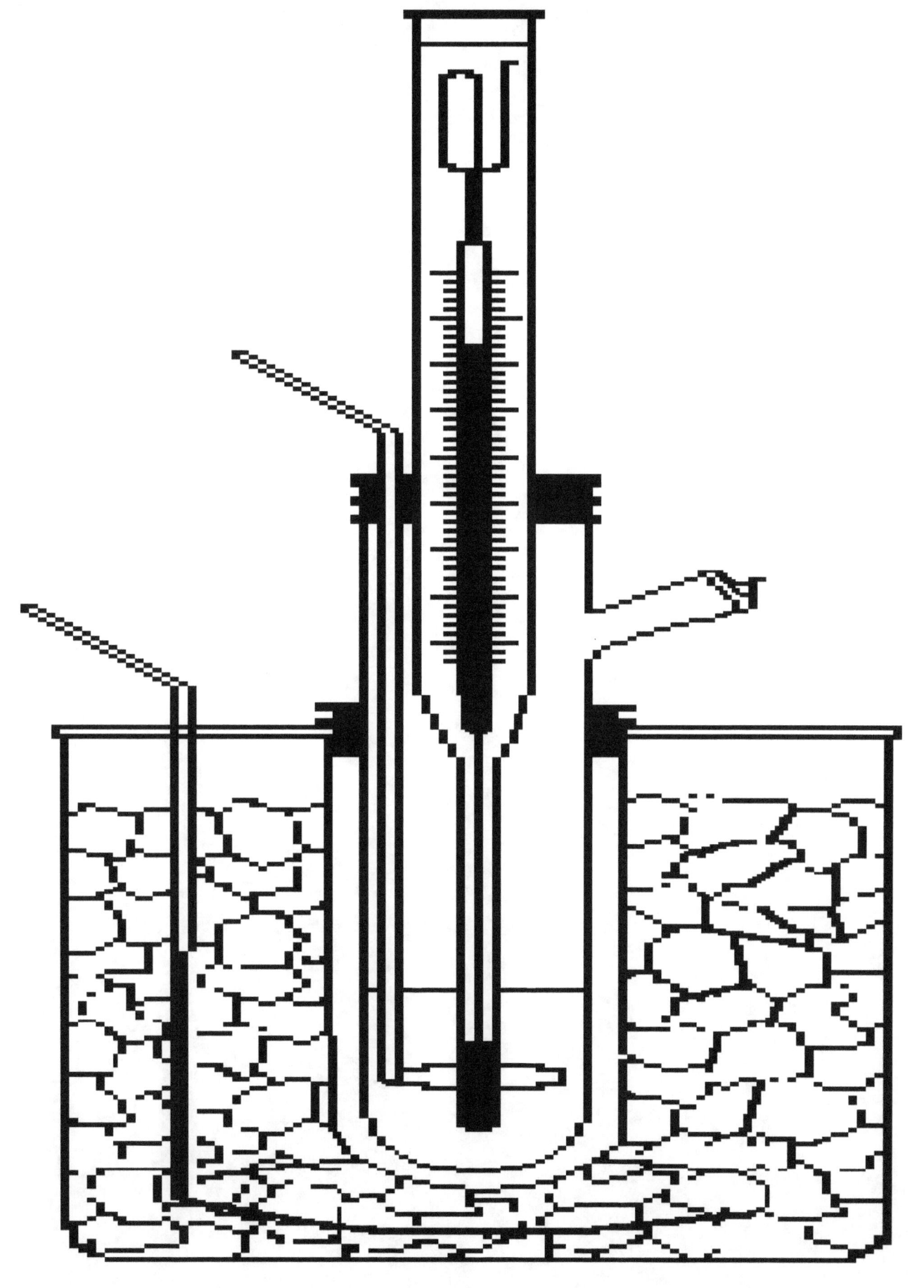

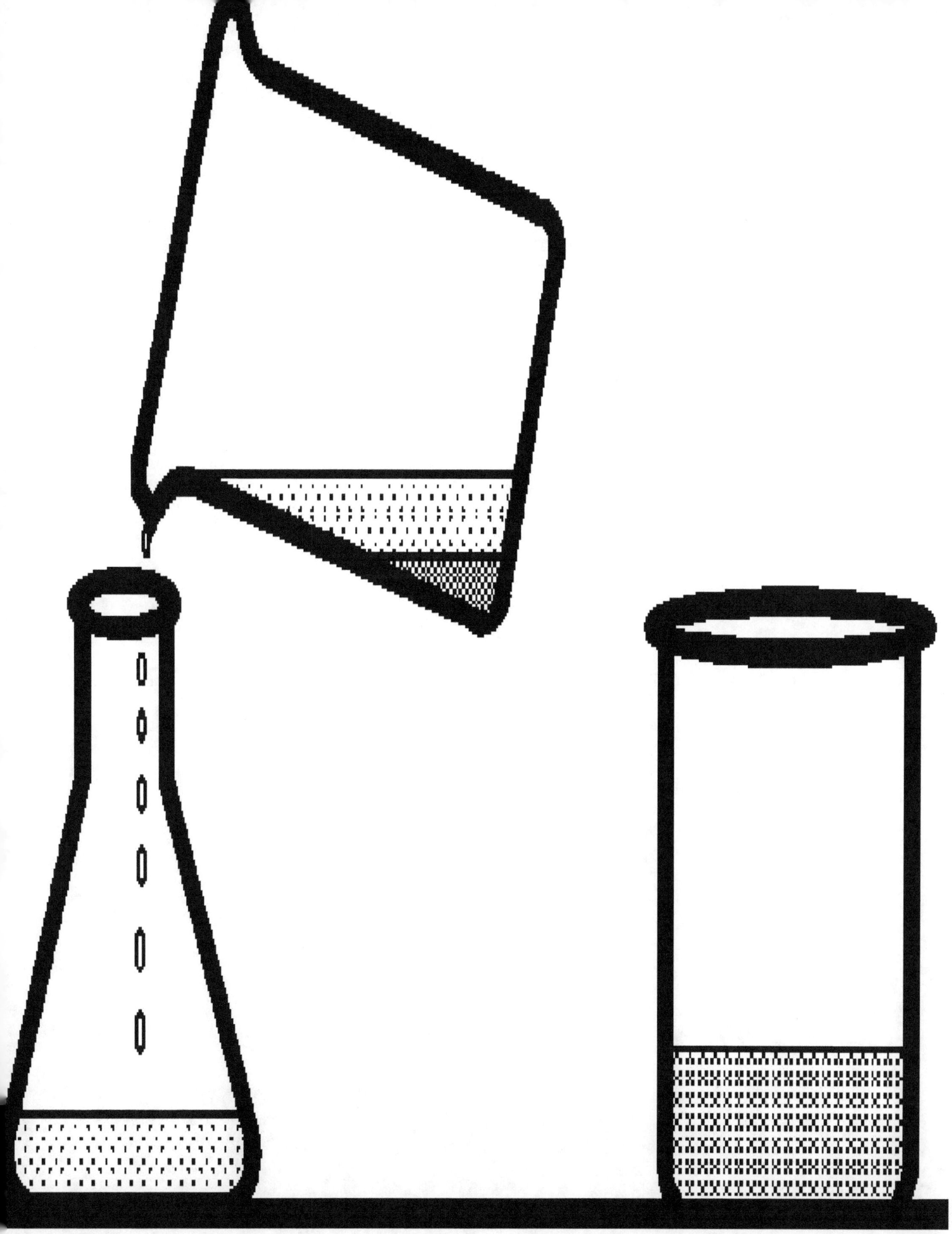

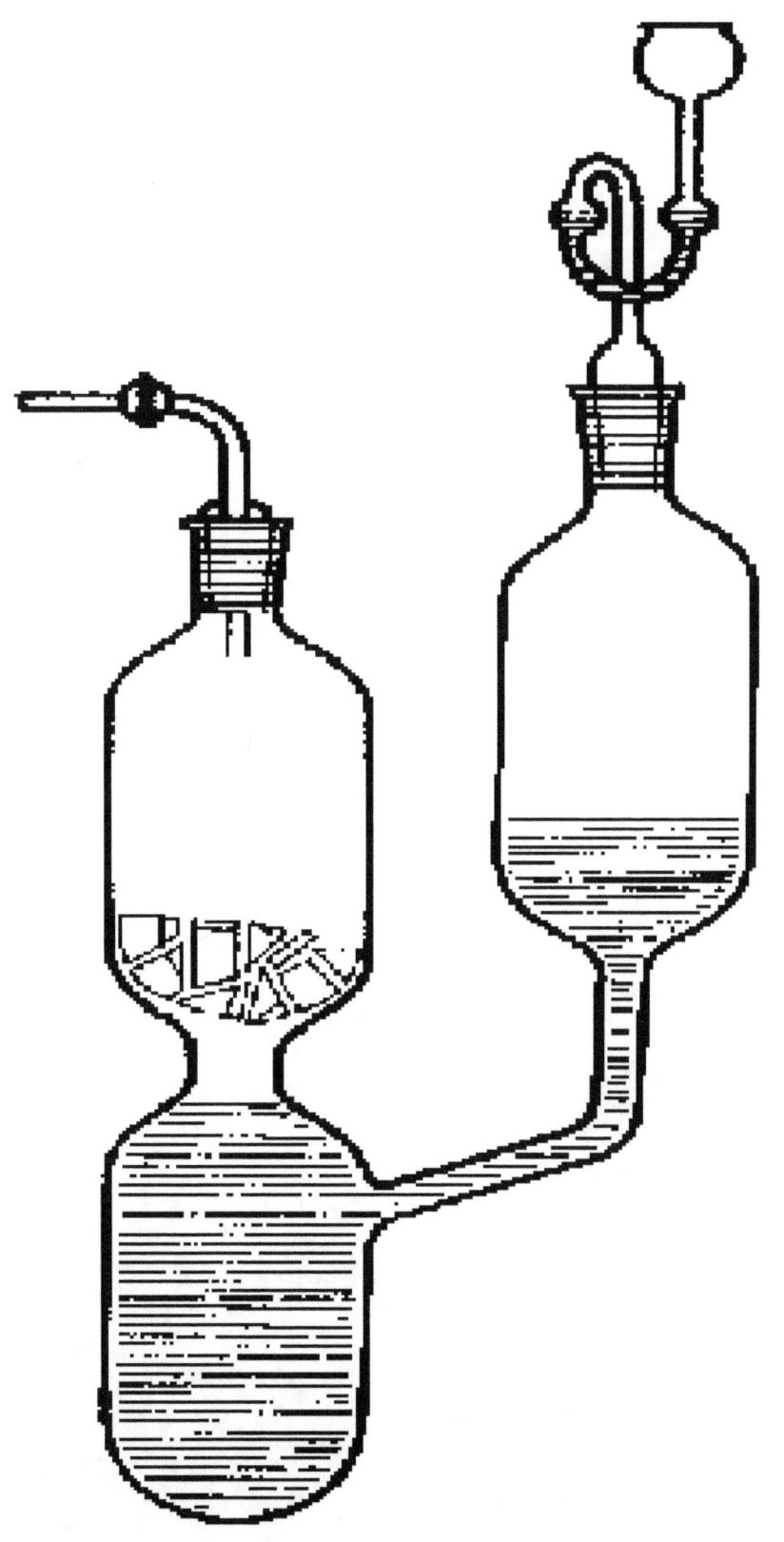

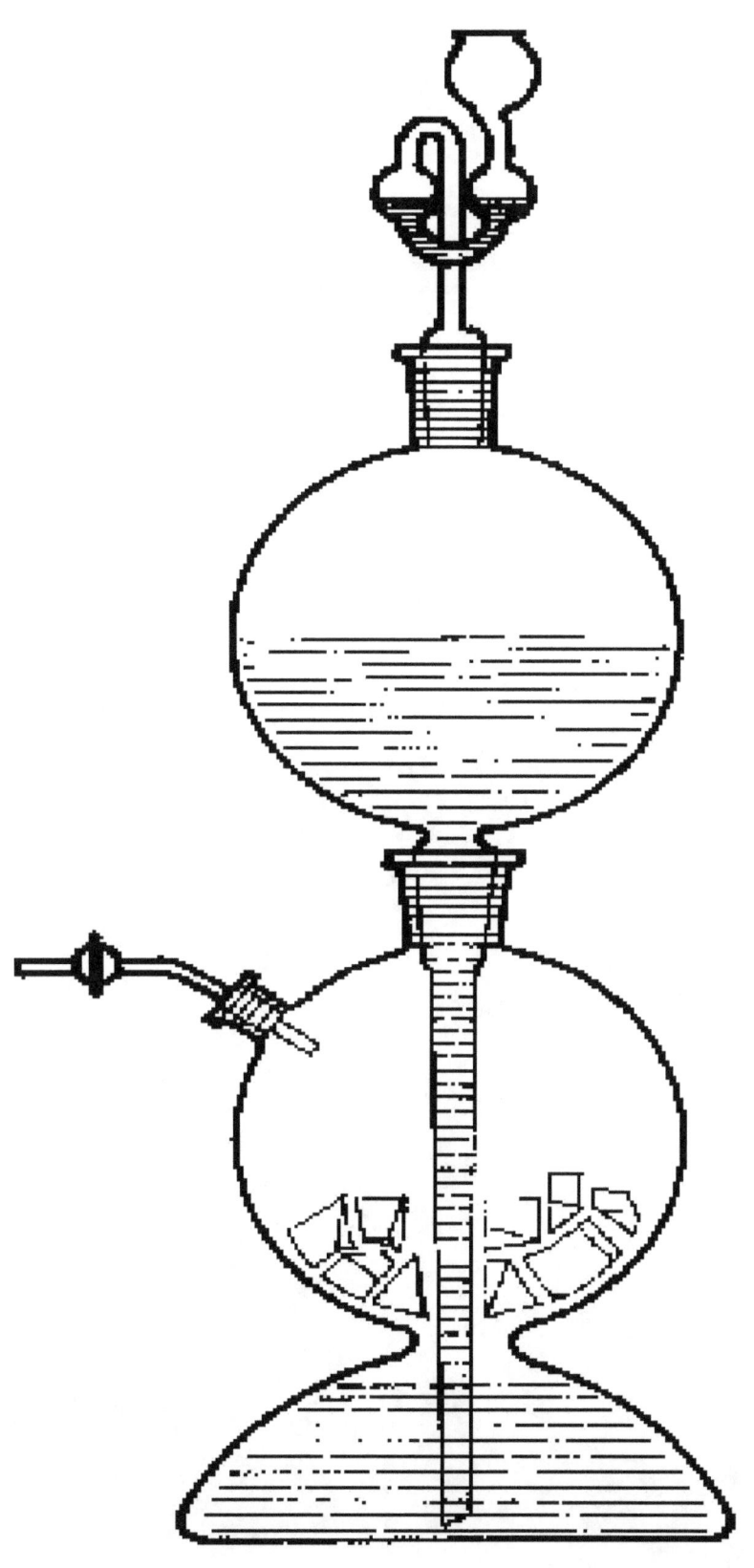

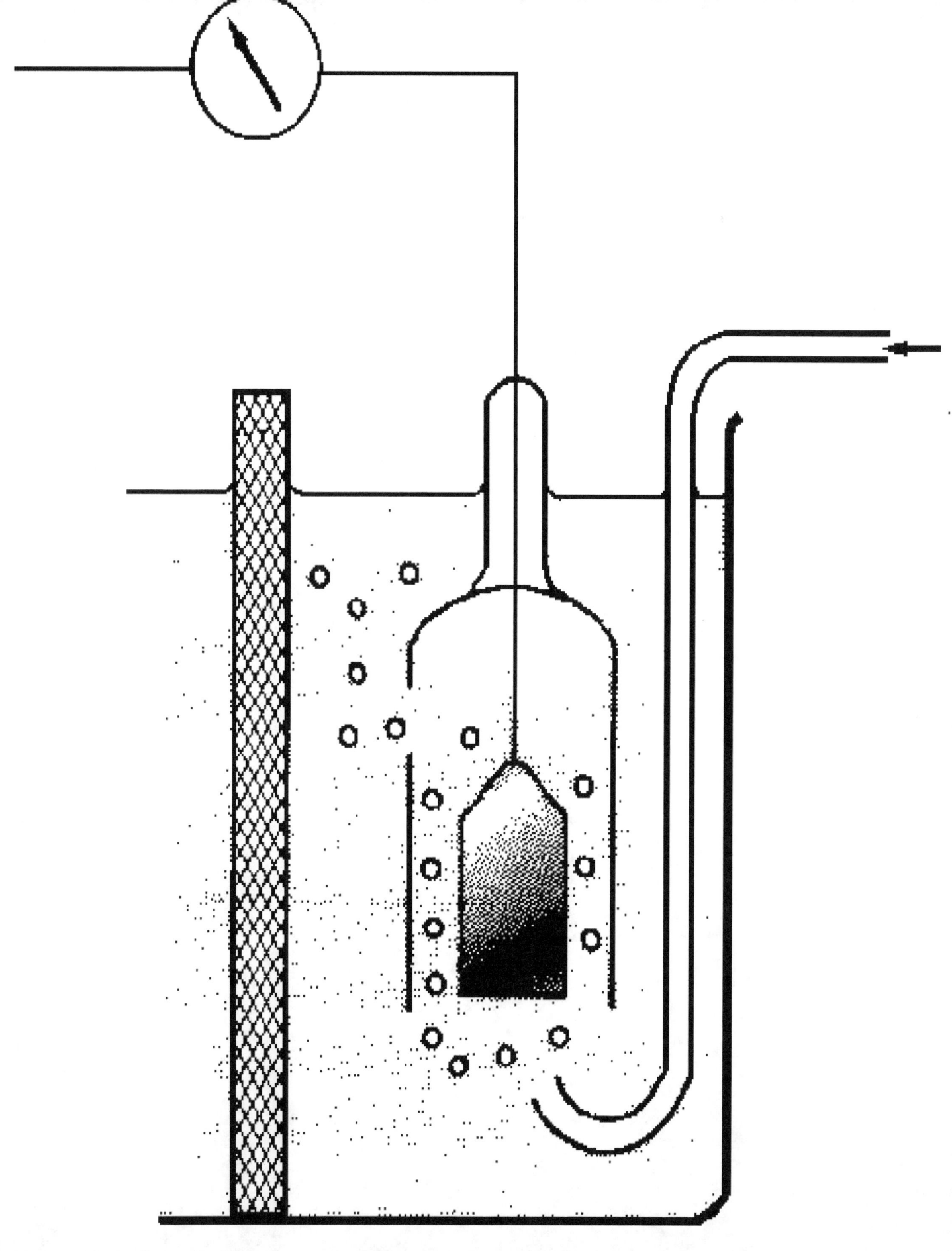

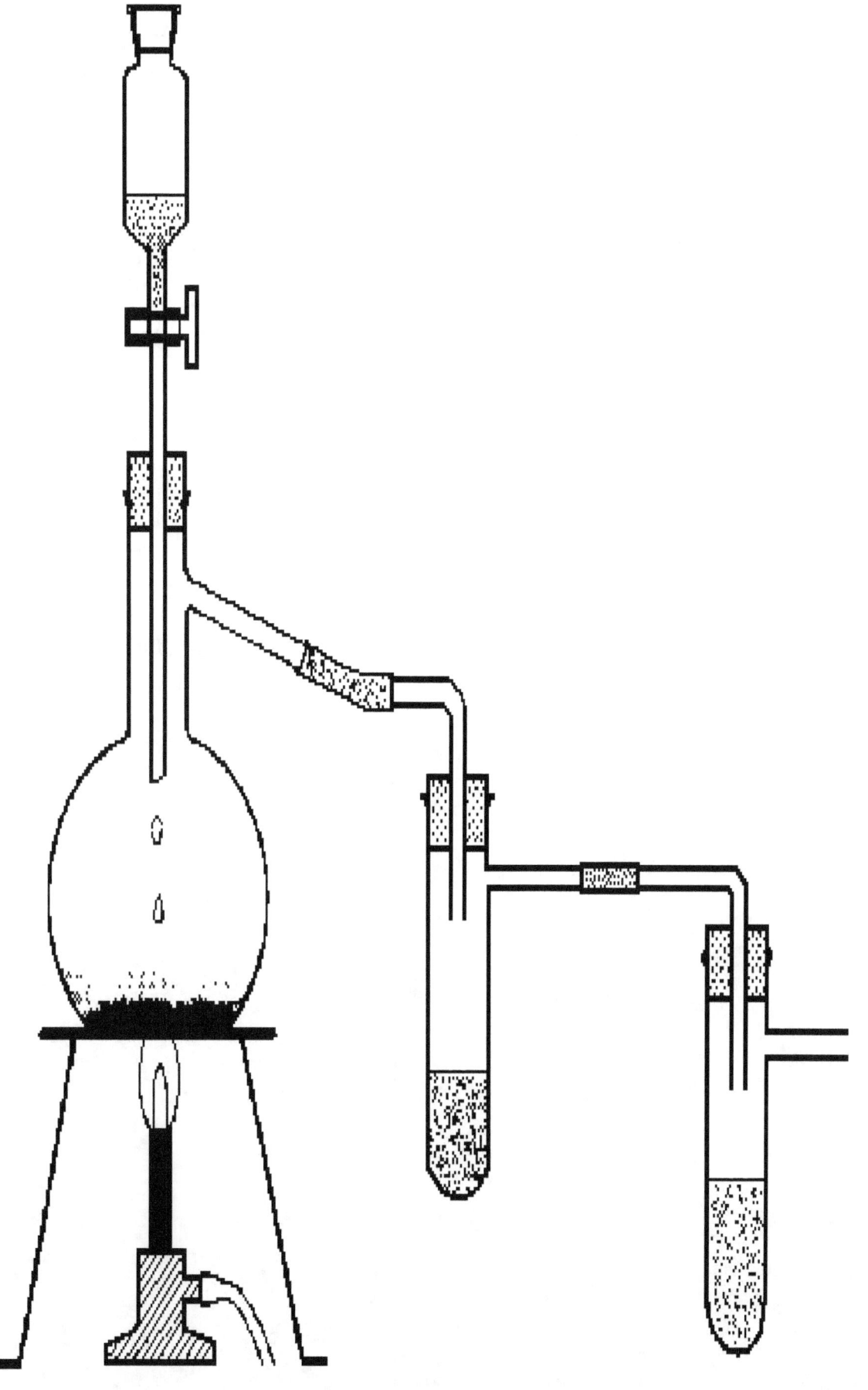

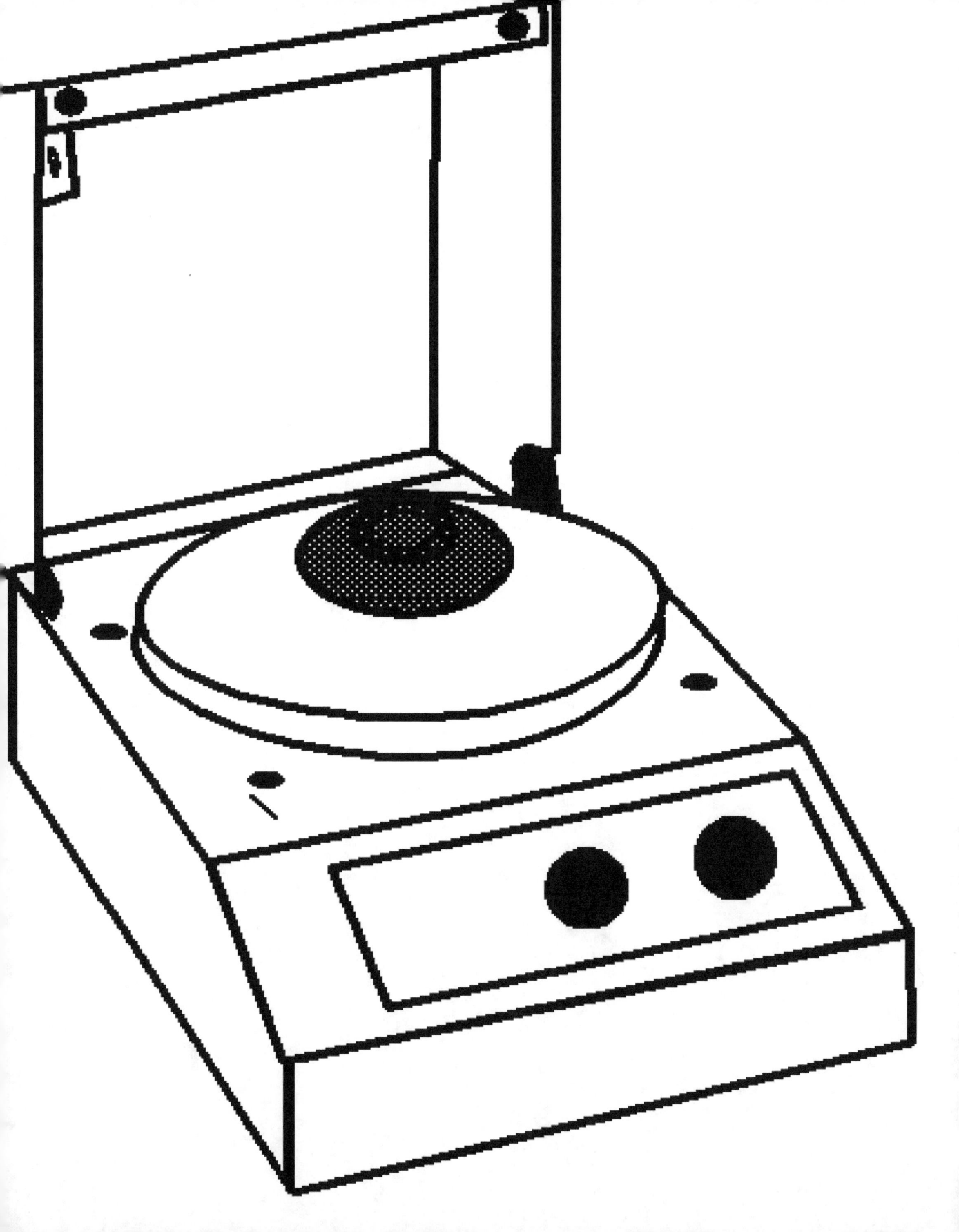

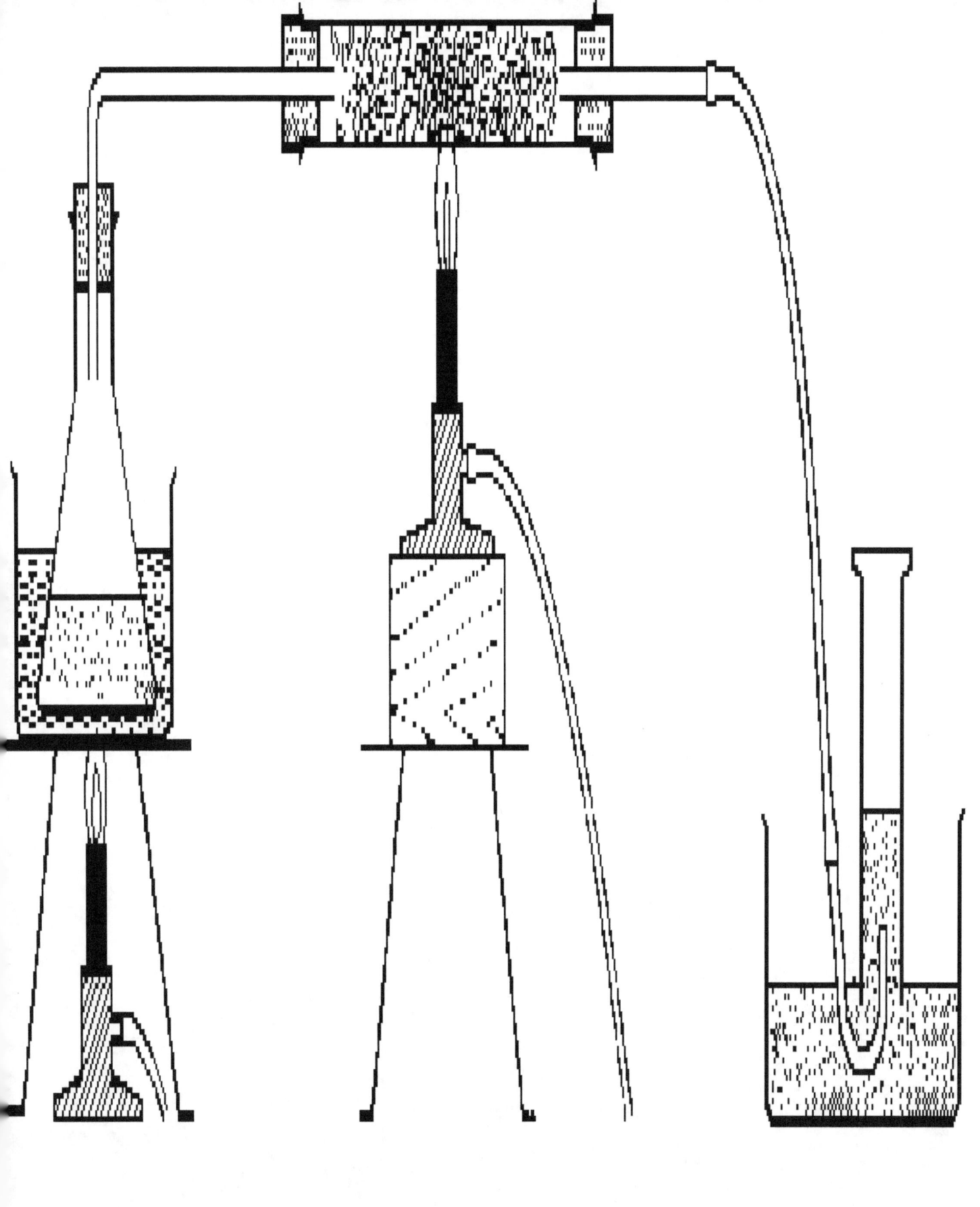

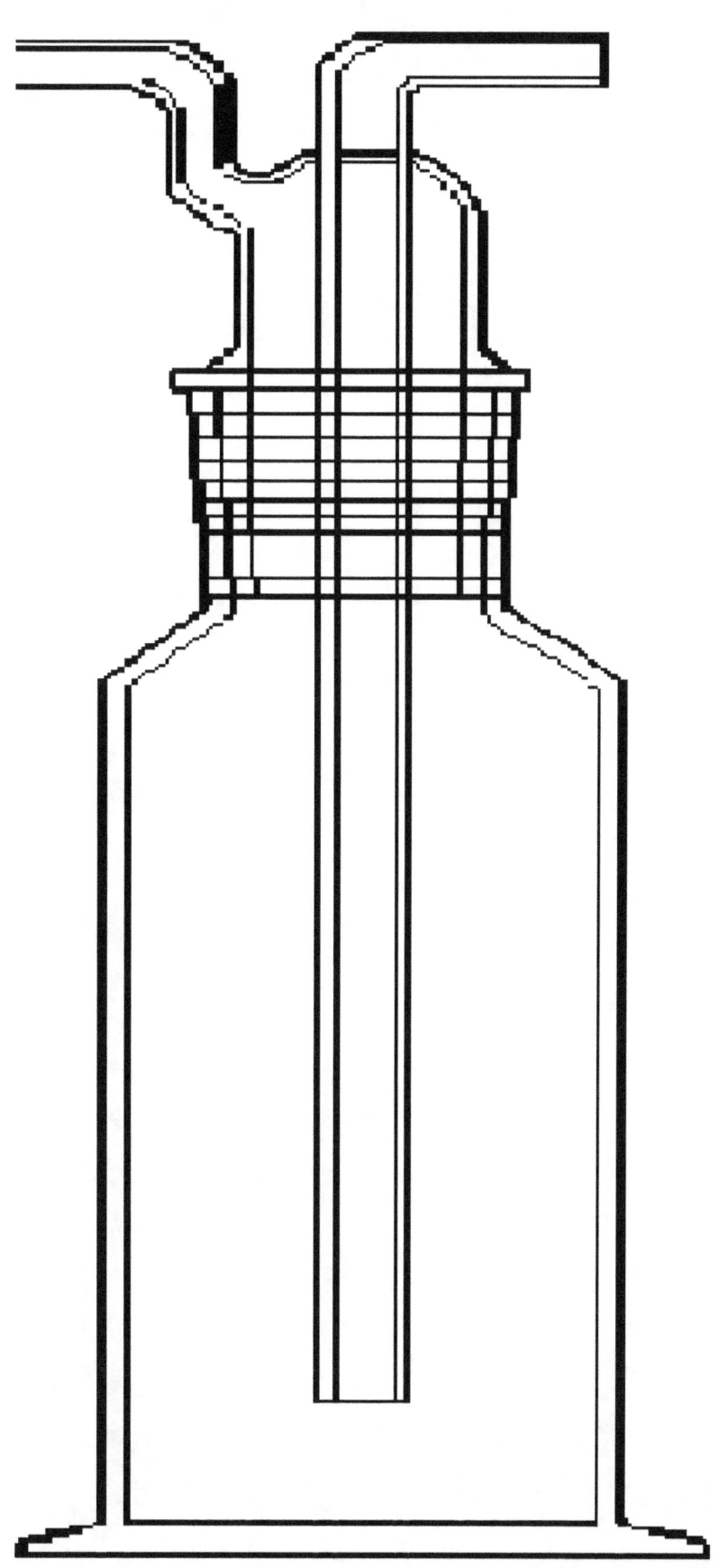

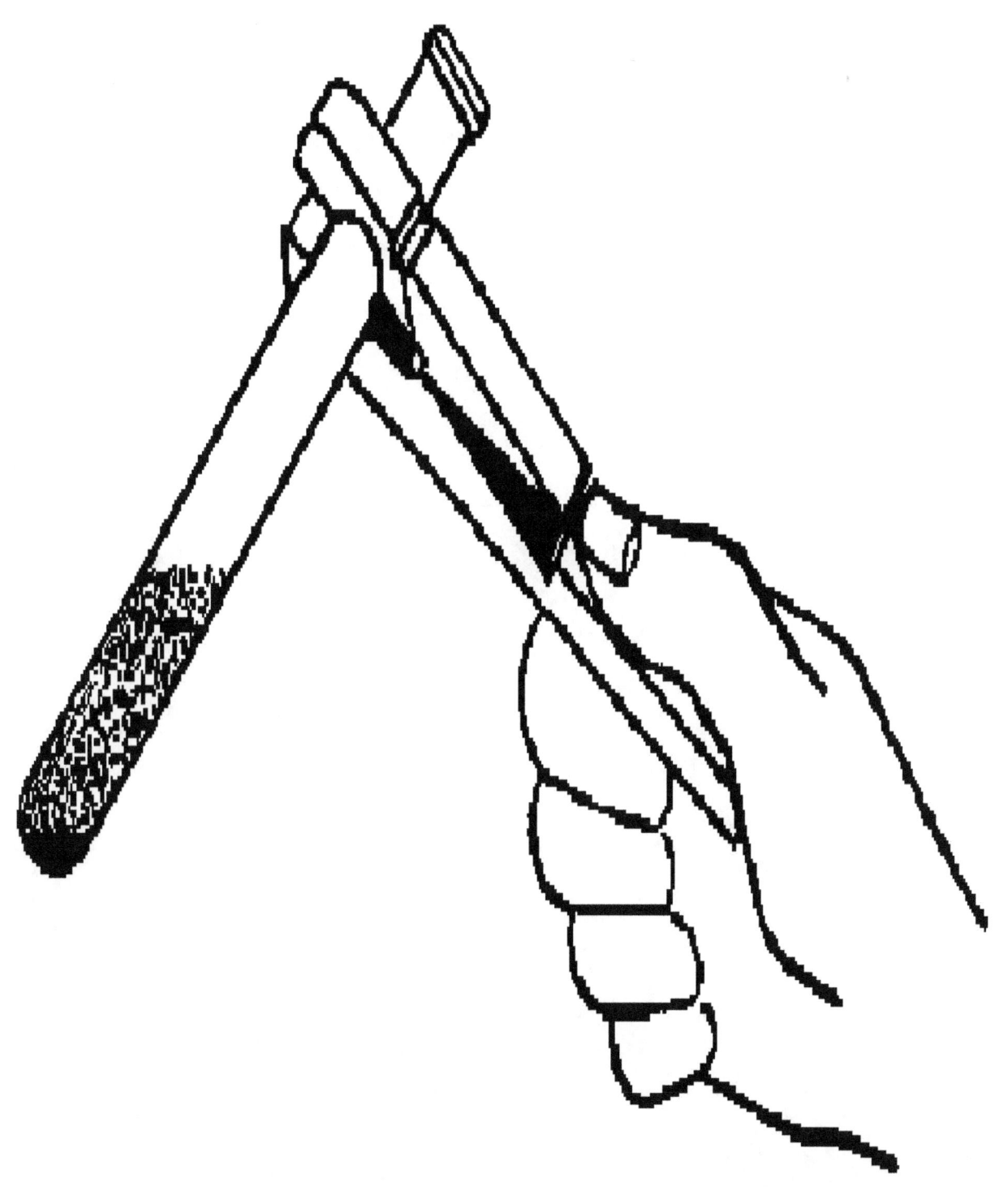

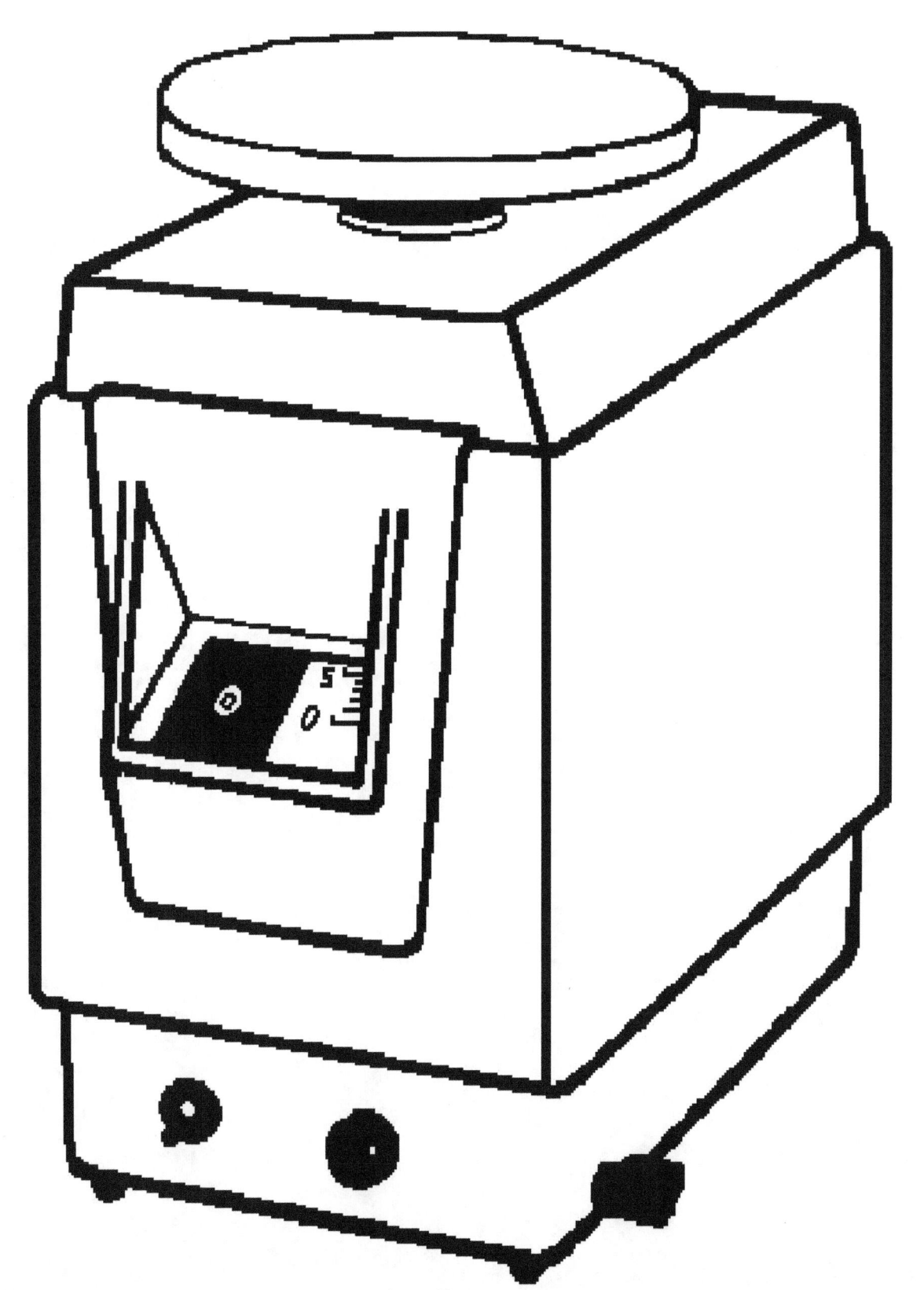

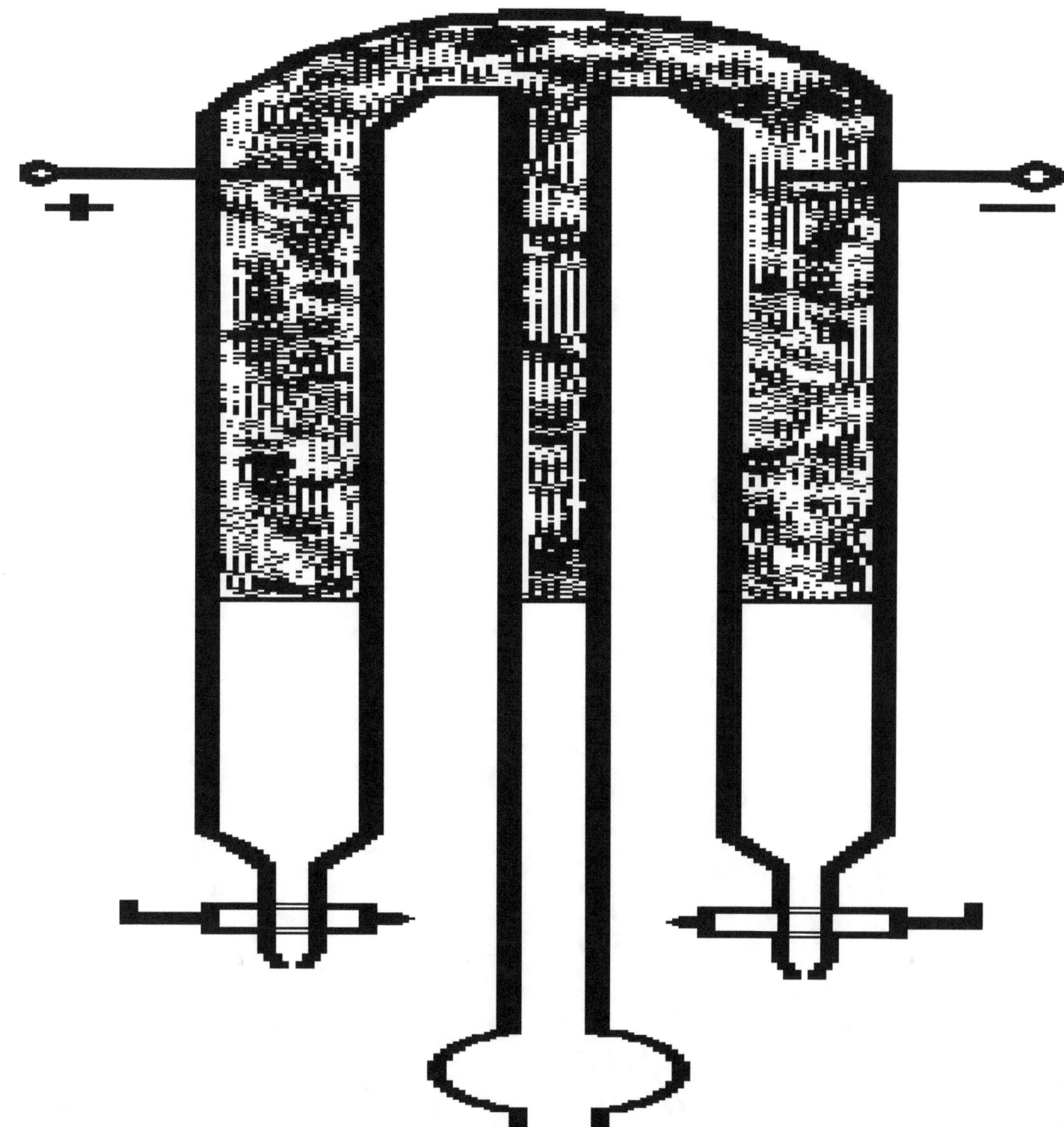

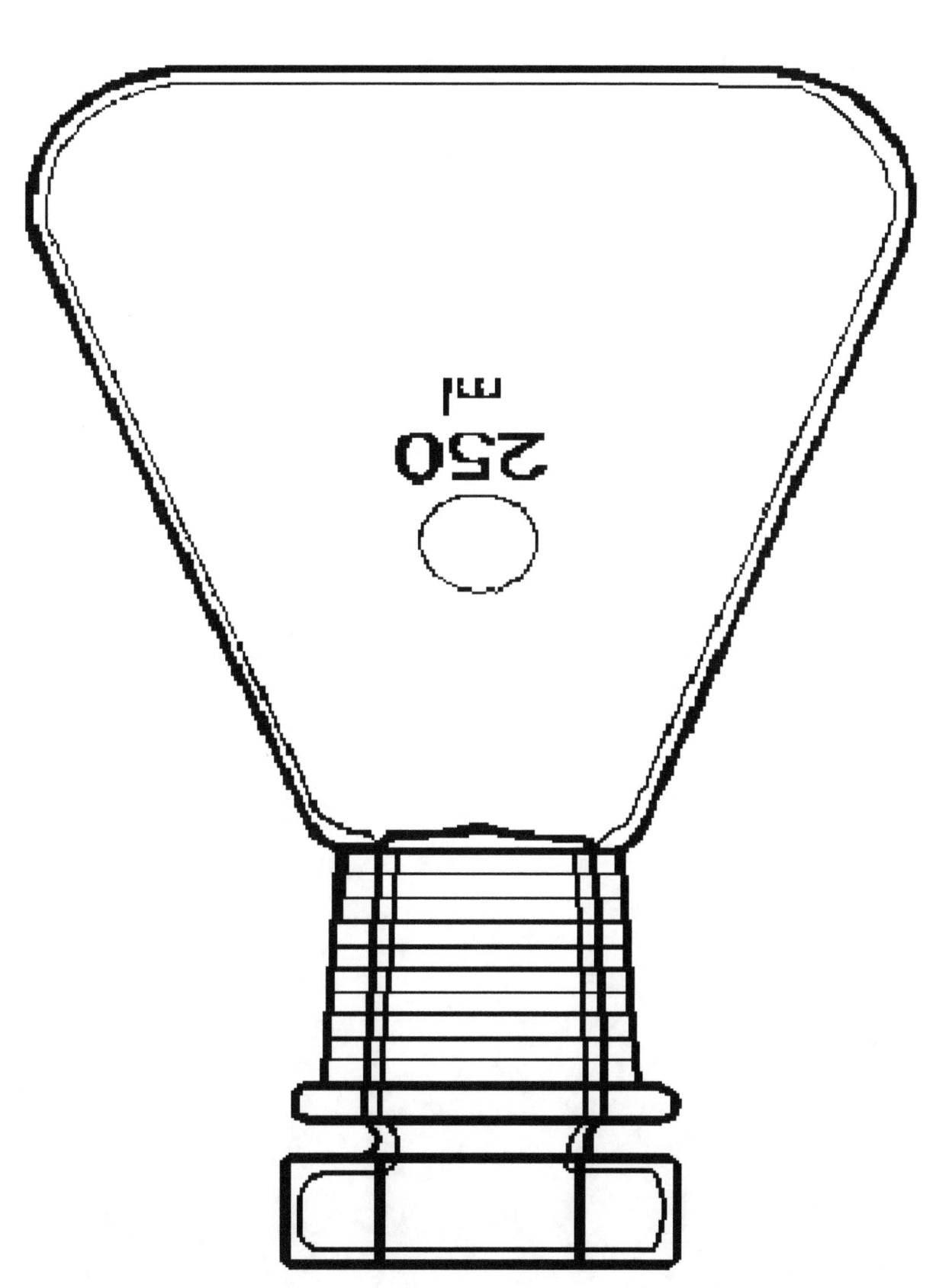

www.ingramcontent.com/pod-product-compliance
Lightning Source LLC
Chambersburg PA
CBHW080528190526
45169CB00008B/3100